BIRDS

Rebecca Woodbury, Ph.D., M.Ed.

Gravitas Publications Inc.

BIRDS

Illustrations: Janet Moneymaker

Birds
ISBN 978-1-950415-66-3

Published by Gravitas Publications Inc.
Imprint: Real Science-4-Kids
www.gravitaspublications.com
www.realscience4kids.com

RS4K

Photo credits: Cover & Title Pg: Frank Winkler, PubDomain; Above: Rebecca Tregear from Pixabay; P.5. Markus De Nitto from Pixabay; P.7. Zdeněk Macháček on Unsplash; P.9. Lowell, AdobeStock; P.11. Joshua Ryder on Unsplash; P.13. Catherine Merlin on Unsplash; P.15: Top, Couleur from Pixabay; Bottom, Akram Huseyn on Unsplash; P. 17. Mabel Amber, who will one day from Pixabay; P.19. Brian, AdobeStock

Birds can be found
all over the world.

Some birds, like **penguins**,
live on the cold **Antarctic** ice.

Why don't their
feet freeze?

Other birds, like the **toucan**,
live in warm tropical climates.

I wonder how they can carry that big beak everywhere.

Strong neck?

All birds have **wings**. Most birds use their wings for flying.

Tiny **hummingbirds**, zip around with fast-moving wings, looking for bugs to eat and nectar to drink from flowers.

Penguins cannot fly. Instead, they use their small wings as flippers for swimming.

The **ostrich** is another bird that cannot fly. But ostriches have long legs for running very fast.

I think I will stay away from that one!

All birds have **feathers** and lay **eggs** with hard shells. Birds have hollow bones to make them lighter so they can fly.

I wonder what it is like to have feathers on your arms.

Maybe then we could fly!

Most birds live on land.
Some birds, like ducks, spend
most of their time in water.

Some birds eat seeds, and some eat insects or small animals. Some birds, like the roadrunner, eat lizards!

Roadrunners can fly.

Yes. But they would rather run!

How many different kinds of birds can you see where you live? How much can you learn about these birds by watching them?

For fun, keep track of your bird observations by writing and drawing them in a notebook.

How to say science words

Antarctic   (ant-AHRK-tik)

bird   (buhrd)

feather   (FEH-thuhr)

hummingbird   (HUH-ming-buhrd)

ostrich   (AW-strich)

penguin   (PEN-gwuhn)

roadrunner   (ROHD-ruh-nuhr)

science   (SIY-uhns)

toucan   (TOO-kaan)

wing   (WING)

www.ingramcontent.com/pod-product-compliance
Lightning Source LLC
Chambersburg PA
CBHW040149200326
41520CB00028B/7538